Hana Chebaani
Nora Abid

Controlling the date moth

Hana Chebaani
Nora Abid

Controlling the date moth

Use of Bacillus thuriengiensis and Diflubenzuron against date moth larvae

ScienciaScripts

Cover image: www.ingimage.com

This book is a translation from the original published under ISBN 978-620-6-71781-2.

Publisher:
Sciencia Scripts
is a trademark of
Dodo Books Indian Ocean Ltd. and OmniScriptum S.R.L publishing group

120 High Road, East Finchley, London, N2 9ED, United Kingdom
Str. Armeneasca 28/1, office 1, Chisinau MD-2012, Republic of Moldova, Europe
Printed at: see last page
ISBN: 978-620-7-92426-4

INTRODUCTION

THE date palm is the fruit tree par excellence of the Saharan desert where he plays A role at a time A role economic grace to the production of the dates Who rich in nutrients, provides a multitude of secondary products, and an ecological role since it gives its structure to the oasis. There area of the Algerian palm groves is estimated has 101,820 Ha (1.2% of there SAU) with a workforce of 12,035,650 palm trees and a production of 253,320 tonnes all varieties combined, located mainly in THE wilaya of Biskra and El-Oued. There production of the Deglet Nour variety is estimated at 97,940 tonnes (38.66%), that of the Ghars variety has 34,630 tons (13.67%) And Finally THE others varieties, affect the 120,750 tonnes (47.66%) (ANONNYMOUS, 2003) However this potential phoenicicole stay very fragile against some diseases and formidable pests (IDDER, 1984). Around fifty species attack the date palm and its products, most of which belong to the class of insects. Some feed on the sap, others consume the sap. flippers and the drink, Finally else se develop in costs of the flowers and green, ripe or stored fruit. He This is Oligonichus afrasiaticus , Parlatoria blanchardi and the moth of the dates (Ectomyelois ceratoniae). There moth of there date Ectomyelois ceratoniae Zeller, It is A lepidoptera Who causes of the problems recurring In THE plots of date palm in THE request and in Algeria. Dates are easily detached from the bunch and are thus lost for harvest. Those infested are depreciated has cause of there presence of lumps excrement and often of the larvae (ANONYMOUS ,1992). THE losses due to attacks of there moth of the dates are estimated between 10 has 40% in the field and can reach 40% or more in stock (ANONYMOUS, 2005). As part of better management of date palm pests, more particularly the moth of the dates, We have door our infestations of a comparative approach to biological and chemical control with a view to better understanding the biological activities of these tested products. The objective of this memoir East to improve THE knowledge fight organic with Bacillus thuringiensis and the struggle chemical with A regulator of growth diflubenzeron and we tried to experiment the latter products against the L3 larval stage of the date moth. For reach the objectives, we have organized This work in two parts theoretical and practical.

- A part theoretical Who understand the data bibliographic on the moth dates, as well as some data concerning the entomopathogen
Bacillus thuringiensis And the insecticide used In our tests.

- A part experimental of which which We have addressed THE material used And methodology adopted, the results and discussion are treated in the last part and finally a general conclusion.

CHAPTER I
DATE MOTH

1. Historical And distribution geographical

Ectomyelois ceratoniae Zeller is a Lepidoptera Pyralidae. It is commonly called the locust bean moth in North Africa. The species was described in 1839 by Zeller. (AGENJO, 1959) Quoted by (DOUNANDJI, nineteen eighty one). According to LEBERRE (1978), the LORD WALSINGHAM, highlighted the presence of Ectomyelois ceratoniae Zeller in Algerian dates in 1904 while JACOBS, noted its presence in dates from the Middle East in 1933. The distribution areas of Ectomyelois ceratoniae in the world correspond to three different types of climate: tropical, Mediterranean and continental, (DOUMANDJI, 1981). Thus this common, cosmopolitan lepidopteran And likely to be found anywhere in the world. Being ecologically eurythermic, its distribution area is very vast, extending from 30th degree south latitude to 50th degree north latitude. (LEPIGRE, 1963 and BALACHOWSKY, 1972). In Algeria, this pest reported two areas of multiplication of Ectomyelois ceratoniae . The first, a coastal border 40 to 80 kilometers wide, extending for nearly 1000 kilometers. The second consists of all the oases, the most important of which are located along the southeast. On the high plateaus, this lepidoptera seems absent at most, it still manages to go down to the north of Médéa and Constantine a few kilometers from these cities (DOUMANDJI.1981).

2. Systematic

According to FAT (1951), Ectomyelois ceratoniae belongs to: Phylum: Arthropoda Below phylum: Mandibulates Class: Insecta Great order: Mecopteroids Order: LepidopteraBelow order : Heteronera Divisions: Dtrysians Group: Heterocera Super family: Pyraloidea Family: Pyralidea Below family: Phycitinae Gender: EctomyeloisSpecies: Ectomyelois ceratoniae

3. Morphology

3.1. The egg

The egg is oval in shape whose largest dimension is 0.6 to 0.8 mm, It is surrounded by a cuticle translucent (DOUMANDJI, nineteen eighty one), he is of White colour clear when laying eggs then acquire a pink color, if it is fertile (DHOUIBI and JARRAYA, 1988).

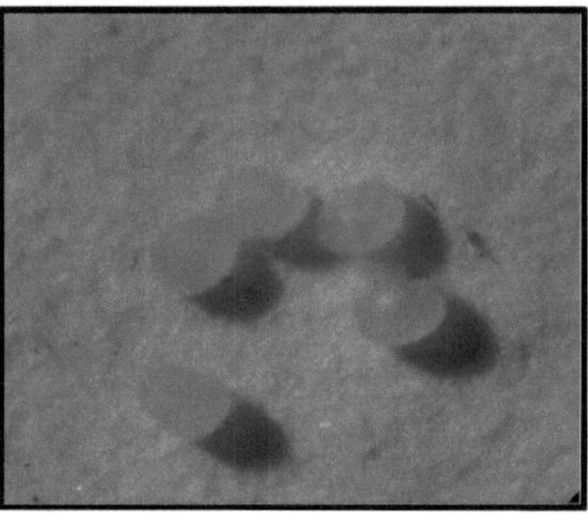

Figure 1 : Eggs of Ectomyelois ceratoniae (Photo Abid And Chebaani) (X40 magnification).

3.2. Larva

Larva is polyphagous, eruciform in shape. It is entirely pink or yellowish white, with head Brown, (DOUMANDJI, nineteen eighty one) And DOUMANDJI-MITICHE ,1977). However DHOUIBI (1989) noted that the color of the larva is yellowish red after hatching, For become YELLOW pinkish by there following, At last stadium, there color of there larva is influenced by his food. Caterpillars feeding on dates are dark pink in color, while those feeding on pistachios and grenades are light pink in color to yellowish. If the food is contaminated by mold (saprophytic fungi, cases of the pomegranates to the harvest). The digestive tract appears black. As pupation approaches, the color intensity decreases and the larvae become yellowish in color. The color of the head capsule, everyone THE stadiums is brown clear at the time of moulting and turns uniformly red dark afterwards . Their body And constituted of 12 segment has leave of segment cephalic, the thoracic segments carry the three pairs of walking legs and the abdominal segments present the four pairs of false legs or suction cups (SAGGOU, 2001). Growth occurs through successive molts during which the length of the tracks increases. The length reached is 18 mm with a width of between 0.1 and 0.3 mm (LEBERRE, 1978). There are five larval stages which differ from each other by size, capsule size cephalic (DHOUIBI and JARRAYA, 1988).

Figure2: Larva of Ectomyelois ceratoniae (Photo Abid And Chebaani) (25x magnification).

Female Male

Figure3: Larva female and male of Ectomyelois ceratoniae (Photo Abid And Chebaani) (25x magnification)

3.3. Chrysalis (nymph)

The nymph measures 9 to 11 mm in length (DHOUIBI, 1991), and has a cylindrical-conical shaped body (DOUMANDJI, 1981). Its testaceous brown chitinous envelope is surrounded by a loose silk sheath woven by the caterpillar before its nymphal moult (LEBERRE, 1978). It presents a sexual dimorphism relating to the location of the virtual genital orifice (^{8th} sternitis in females and ^{9th} in the male), characterized by the presence of seven pairs of spines on the first seven abdominal segments and two hooks at the abdominal end (curl downward). The prothorax is rough with an irregular midodorsal keel, characterizing the pupa of Ectomyelois ceratoniae (DHOUIBI and JARRAYA, 1988).

3.4. Adult

The adult of the date moth is a small butterfly with twilight and nocturnal activity, it is light gray in color. The front wings are decorated with more or less marked designs and are relatively narrow; widening slightly towards the end. The hindwings uniformly pale and bordered by a whitish silky row. Females have a larger wingspan than males: body length varies from 6 to 12 mm, the wingspan of 16 to 22 mm (BEN AYED, 2006). The sexual dimorphism is very clear at the level of the posterior end of the abdomen, thus facilitating the recognition of the sexes with the naked eye in the female, there is at the level of the abdominal end a small, relatively dark depression from which emerges intermittently a sclerotized and retractile organ corresponding to the ovipositor. He is fusiform and ends with a light-colored hairy appendage, corresponding to anal papillae, at the end of which the egg-laying orifice opens: this is the Ostium oviductum . In the male, at the level of the genital frame two valves are visible externally and which allow easily sex adults (DHOUIBI, 1989).

Figure 4 : Adult of Ectomyelois ceratoniae (Photo Abid and Chebaani) (X25 magnification).

4. Cycle biological

Ectomyelois ceratoniae is a microlepidoptera, which completes its biological cycle through the passage of different stages: egg, caterpillar, nymph, adult. THE emergences have place In there first part of there night. According to (WERTHEIMER ,1958), cited by (IDDER, 1984). Butterflies mate in the open air or even inside the enclosure where they were born. The relatively long copulation lasts several hours, and a single mating is sufficient for oviposition. A female lays 60 to 120 eggs arranged singly or in small, loose groups. Egg incubation time varies between 3 and 7 days depending on temperature. Right away after hatching, the caterpillar searches for food and a shelter. She drills holes and hollow a gallery and locates between the pulp and nucleus (VILARDIBO, 1975). According to (WERTHEIMER, 1958) cited by (IDDER, 1984), most often the caterpillar author weaves from it a cocoon with very loose meshes which it can quickly leave towards the front or towards the back as soon as it is disturbed, or simply to feed itself, this is the old winter caterpillar stage which constitutes the resting or waiting stage. During this time, nutritional and motor functions are slowed down. The growth of caterpillars as a function of ambient temperature from 6 weeks to 8 months (VILARDIBO, 1975). There chrysalis settles down In there date, there head always tour towards THE hole communication with the outside world (IDDER, 1984). According to (LEPIGRE, 1963), there pupation has a indefinite period. The imago which results in a shelf life of 3 to 5 days during which, it will mate and lay eggs.

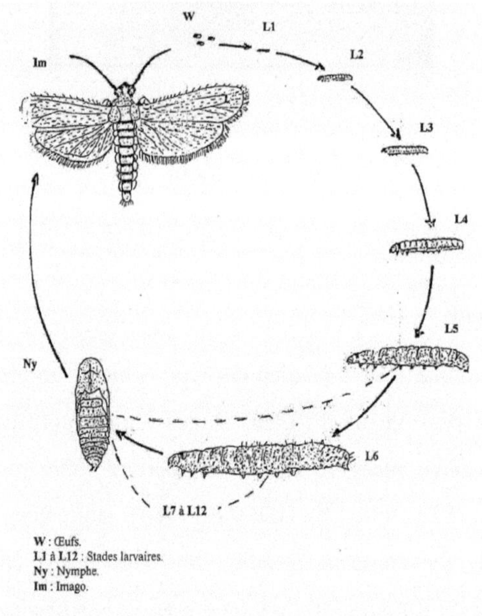

Figure 5: Cycle biologique de la pyrale de la datte (DOUMANDJI, 1983).

5. Number of generation

Ectomyelois ceratoniae is a polyvoltine species, in which under good conditions, four generations can succeed one another during the year. But in fact this number of generations varies from 1 to 4 depending on climatic conditions and host plants (DOUMANDJI, 1981). According to (LEPIGRE, 1963) first exit of the adults spreads of the end March in mid april.

- THE second begins THE June 15 And se continues to surroundings of 20

August.

➢ THE third flight extends of the last days from August until at the end October, beginning of November: this is the most formidable generation and responsible for the main damage to dates.

➢ THE fourth flight intervenes at there END of November. This generation restraint is superimposed in time on the third.

Table1: Number of generations of Ectomyelois ceratoniae (WERTHEIMER, 1975)

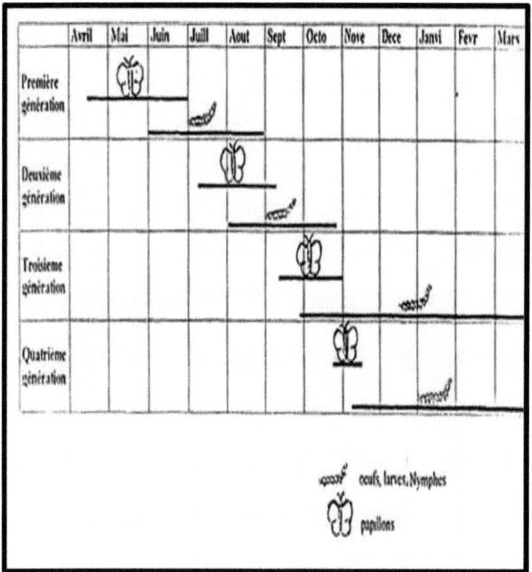

6. Plants hosts

Ectomyelois ceratoniae is a species very polyphagous, she also lives GOOD on fruit ripe or close to maturity, as in dried fruits stored in warehouses and stores (BEN AYED,2006). (DOUMANDJI, 1981) reports 49 species of host plants in the world, 32 in Algeria including 25 in Mitidja. The main host plants in Algeria are the carob tree (Ceratonia siliqua. L) , THE medlar of Japan (Eriobotrya japonica) , the orange tree (Citrus sinensis L) , THE pomegranate (Punica granatum.L) and date palms (Phoenix dactylifera.L) . Then come the secondaries represented by Acacia farnesiana , R'tem Retama bovei , then the occasional plants, we cite the almond tree (Prunus amygdalus batch), the apricot (Prunus armeniaca.L), the apple tree (Malus pumila . Miller), figs (Ficus carica .L).

7. Damage caused

According to BOUKA et al (2001), and ANONYMOUS (2005), losses due to attacks of the moth of the dates are estimated between 10 has 40% At field and can reach 40% Or no longer in stock. The attacked dates are easily detached from the bunch and are thus lost for the harvest. Those infested are depreciated has cause of there presence lumps excrement and often larvae (ANONYMOUS, 1992). At stock THE dates can be completely degraded until to nuclei by the caterpillars (BLACHOWSKY, 1972).

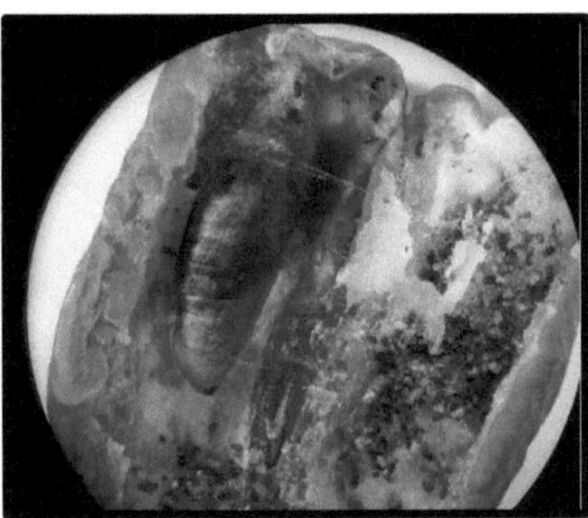

Figure 6 : Damage caused by of Ectomyelois ceratoniae on the date (Photo Abid and Chebaani).

Figure 7: Damage caused by Ectomyelois ceratoniae on grenadiers (DHOUIBI,1991)

8. Means of struggle

In the context of sustainable development, and in order to achieve more phoeniciculture respectful of the environment and of there health human, in aiming to reduce the damage caused by the date moth while calling for a reasoned fight.

8.1. Struggle cultural

There struggle cultural constitutes there main measure preventive aimed at decrease of level of the insect population, it translates as follows:

❖ Pickup of the waste of the fruits hosts (grenades, apricots And dates…) And their cremation For ensure there reduction of the sources of food and the destruction of hibernation sites (DHOUIBI, 1989).

❖ There size of the djerids, of the cornafs And of the diets No harvested.

❖ Cleaning of diets of the dates And kidnapping of the infested fruit of the diets.

❖ Bagging bunches is an operation carried out regularly every year in oases can constitute an effective solution which both reduces losses caused by rain and limits fruit moth attacks (DHOUIBI, 2000).

❖ Treatment of the deposits Before THE beginning of there campaign.

❖ Cleaning of local in THE basting with of there lime long live And in THE treating has base of an insecticide.

8.2. Struggle chemical

In Tunisia, since 1974, the only means used against locust bean moths is chemical control. This may have reduced populations of the pest living on date palms only. This involves ensuring that the fruits are covered by a insecticide to poison the eggs and young larvae shortly after hatching and before penetrating the fruit (DHOUIBI, 2000).On THE plan economic, the treatments chemical do not have given big satisfaction as long as the infection rate of dates is not negligible when we know that a good part of the date harvest is intended for export which requires a disease-free product and of first quality. In Tunisia, the use of Deltametrine made it possible to reduce the percentage of infestations from 24 to 17%. Among the other powdered insecticides used are MALATHION 2%, PARATHION 1.25% and PHASALON 4% (DHOUIBI, 1989). Currently, the products used in chemical interventions carried out by the services of the protection Plant we quoted on DIMILIN 2%, ASMIDON3%, the Methyl BROMIDE and PHOSPHINE.

8.3. Struggle Biological

8.3.1. Use insects auxiliaries

The main parasites of Ectomyelois ceratoniae belong to the order Hymenoptera we cite:

- Trichogramma embryophagum Hartig : belongs to the Trichogammatidas family, the length between 0.33 mm and 0.63 mm characterized by a stocky body and a coloring that varies from bright yellow to dark brown.

• It's a ooparasite whose embryonic, laval and nymphal development takes place in the host egg. The imago comes out by cutting a hole of sorts with its mandibles (IDDER, 1984).

• Bracon hebebetor Say: it is a laval ectoparasite belonging to the family Braconidae, 3 mm long, yellowish in color with black areas on the prothorax (DOUMANDJI-MITICHE and DOUMANDJI, 1993). The female of this parasite will bite the host caterpillar, paralyzing it.

After which, the female parasite will lay several eggs on the insect. All larval development of the parasite takes place on the body of the host. At the end of its development, the parasitic larva leaves the larval remains to build its cocoon. pupate not far from the host.

• Phanerotoma Flavitestacea Fischer: is a parasite ovo-larval belongs to the family Braconidae, the laying se done in the bud of the host, larval development of parasite first unfolds from the egg then in different larval stages of the host caterpillar (DOUMANDJI-MITICHE and DOUMANDJI, 1993).

8.3.2. Use of the micro- organisms

Use biopesticides based on bacteria have the advantage of being very specific, non-toxic to warm-blooded animals and other predatory, parasitic and pollinators, as they do not have a phytotoxic effect. The first bioinsecticide, manufactured on an industrial scale, is based on Bacillus thuringiensis . The product is applied by spraying to trees in the same way as an insecticide after ingestion, the bacteria acts on the larvae of the moth in the midgut where the toxic fractions attack the epithelial cells (DHOUIBI, 1991). Other mushroom-based products are marketed, such as spinosad. It is a natural product, extracted from the fermentation of actinomycete (Sacharopolyspora spinosa), he acts by contact between other, on THE stadiums larvae of the lepidoptera. Efficiency of this product was tested against the date moth in Tunisia (KHOUALDIA et al, 2001).

8.4. Struggle biotechnology

8.4.1. There Struggle autocide (THE let go of the insects sterile)

The autocidal or genetic fight against the date moth consists of a massive multiplication of this insect within the framework of artificial breeding. Its sterility is subsequently induced by the use of gamma radiation (TIS), then releases in sufficient quantities are carried out in the oasis palm groves (DHOUIBI, 2000). The ratio of sterile males to normal males increases, triggering a sort of chain reaction, the effectiveness of which increases from generation to generation. The results obtained in the fight against the date worm are very encouraging (DHOUIBI, 2000).

8.4.2. Use of the pheromones

Pheromone traps are used for the purpose of determining the species of insect pest present in a crop, as well as additional investigations or plant protection measures that may be necessary to avoid excessive damage to the harvest (ANONYMOUS, 2006).She consists has disorient the males in diffusing in the atmosphere of there palmerais significant quantities of sexual pheromones, thus preventing the meeting of the sexes and consequently their reproduction which could be promising (DHOUIBI, 1989).

CHAPTER II
GENERAL OVERVIEW OF THE ENTHOMOPATHOGENIC STRAIN AND INSECTICIDE USED

Introduction

He exist of many methods of struggle against THE diseases and the pests pests of crops among these control methods, biological control and chemical control. THE term biological control is reserved for control methods that consist of introducing into a culture A enemy natural, imported else culture Or from massive breeding to combat a well-identified pest. The use of preparations based on micro-organisms such as bacteria, viruses, fungi, called biopesticides is qualified of struggle microphone- biological (JEROME And al , 2002).THE studies more conclusive are those Who have successful at the setting At point of preparation based of bacteria of which some are Today widely broadcast, It is THE case notably preparations containing Bacillus thuringiensis , they are used against several pests (Pierre, 2002).

1. Historical of use of the entomopathogens in struggle biological

The idea of attacking harmful insects by introducing diseases into their population developed to there END of century last (DOUMANDJI, And DOUMANDJI-MITICHE, 1993). It is in 1834 that the Italian BASSI Agostino watch experimentally that Silkworm muscardine is caused by a fungus. In 1865 PASTOR study the Pebrin and flair of worm has silk. There Pebrin is a micro-organism transmitted by the leaves mulberry contaminate Also by the eggs filed by butterflies. In 1874, PASTOR and the county MANCENT have got the idea of combat of the harmful insects through diseases to which they are susceptible. THE first tests have summer facts by METCHNIKOFF in 1879, in Russia. It uses the green Muscardine fungus (Metarrhizium anisopliae) against a grain beetle and a beet weevil. 50 kg of spores of this mushroom are produced. About ten years ago, 2 fungi, 3 viruses and 4 bacteria entered In THE domain of there practical phytosanitary and are marketed In certain countries (REMAUDIER, 1976) cited by (DOUMANDJI-MITICHE and DOUMANDJI, 1993), but multiple difficulties are encountered linked to the acquisition of basic knowledge (identification of the pathogen, its biology, its ecology and its specificity) and the manufacture of the pathogen. Before using these germs, he se revealed necessary of TO DO THE diagostic diseases and the identification of the responsible germs. We must also isolate and cultivate the agents pathogens be sure medium artificial be sure of the organisms living. Finally, We must ensure the safety of pathogens towards vertebrates.

2. General on THE bacteria in struggle biological

THE bacteria are organelles very small, of a few thousandths of mm, more bigger than viruses, they are formed of a cell single, wrapped of a membrane. Their general shape is varied, giving different names to these bacteria (LHOSTE, 1979) cited by (DOUMANDJI-MITICHE, DOUMANDJI, 1993):

- Vibrations: bacteria in shape comma .
- Spirilla: bacteria in shape propeller.
- Bacilli: bacteria in shape of sticks.

Bacteria are found in soil, water, air and in the organism of living beings. PASTOR was the first to realize the importance of bacterial diseases at the insects and the possibility of their use in struggle against THE harmful insects .
THE bacteria include two groups distinct THE bacteria spore-formers having a resistance or sporulated form And non-sporogenic bacteria generally found in the digestive tract and presenting less practical interest than the sporogenic forms due to their very great sensitivity to external agents, in particular to light, drought and contact with air (BALACHOWSKY, 1951).

3. THE bacteria entomopathogens

Bacteria entomopathogens se generally found in order of the Eubacteries, and more particularly the Bacillaceae, Entrobacteriaceae , Micrococcaceae , Pseudomonaceae families . They belong particularly to the genera Serratia and especially Bacillus with the 4 species now well known, Bacillus popilliae , B.moritai , Bacillus Sphaericus And Bacillus thuringiensis . isolated by ISHITAWA of a breeding of towards silk in 1905 in Japan, the latter was forgotten then rediscovered in 1911 by Berliner in Thuringia, and his exploitation agronomic born was considered that in 1950. B. thuringiensis is undoubtedly today the most marketed biopesticide in the world (RIBA and SILVY, 1989) .

4. Bacillus thuringiensis

4.1. Morphology

Bacillus thuringiensis is a bacterium gram positive aerobic and sporulated, we THE find in virtually all soils, water, air and plant foliage, in the shape of a 5mm stick long by 1 mm wide and provided with flagella (BYE and DESCOINS, 1991). Has the ability to kill insects this pathogenic effect is due to protein crystals that the bacteria synthesizes when, in unfavorable conditions, it produces spores (forms of resistance). Ingested by the insect, these crystals release toxins that destroy the cells of its digestive tract, rapid cessation of food consumption and then death (DAHBIA, 1989). He exists of many varieties of Bacillus thuringiensis , each not being toxic that For A number very limited number of entomological species.

4.2. Systematic

- Reign: Protists
- Below reign: Prokaryotes
- Class: Schizomycete.
- Order: Eubacteria
- Family: Bacillaceae.
- Gender: Bacillus
- Species: Bacillus thuringiensis

4-3- Use

The entomopathogenic bacterium Bacillus thuringiensis was the first homologous microorganism In THE world as biopesticide (but THE first has to have summer marketed is the Lemoultine has base mushroom Isaria dense , against white worms, beginning of 20th century). The first approvals date from the 1960s in the United States and the 1970s in France. Preparations based on Bacillus thuringiensis .Concern almost 90% of the biopesticide market, because this bacteria multiplies easily in fermenters and is at the origin of products formulated stable (JEROME et al.,2002). According to (DOUMANDJI-MITICHE, DOUMANDJI, 1993) more than 150 species insects are sensitive At Bacillus thuringiensis .According to (JEROME et al , 2002), this species belonging to three orders; Lipedoptera, Coleoptera, and Diptera. THE Bacillus thuringiensis is easy has obtain industrially and it exist of the units of productions in the USA and USSR in Eastern Europe and France. THE Bacillus thuringiensis has is the subject of a preparation commercialized below THE name of bactospein.

4-4-The toxins of Bacillus thuringiensis

Bacillus thuringiensis toxin often called simply Bt is commonly used as insecticide biological. Of the preparations commercial are available to eradicate caterpillars and other insect pests (JEROME et al , 2002).Bacillus thuringiensis acts on THE insects by the intermediary of toxins main: The endotoxin largely included in a parasporal bipyramidal crital and the heat-stable exotoxin specific to certain strains. This one has a structure comparable to that of a nucleotide which, in addition to its insecticidal effects, gives it mutagenic properties. It is therefore excluded from commercial preparations (RIBA and SILVY, 1989) . According to SEBESTA et al (1981), the exotoxin is released into thc extracellular environment by Bacillus species . Thuringiensis during of the exponential growth phase.

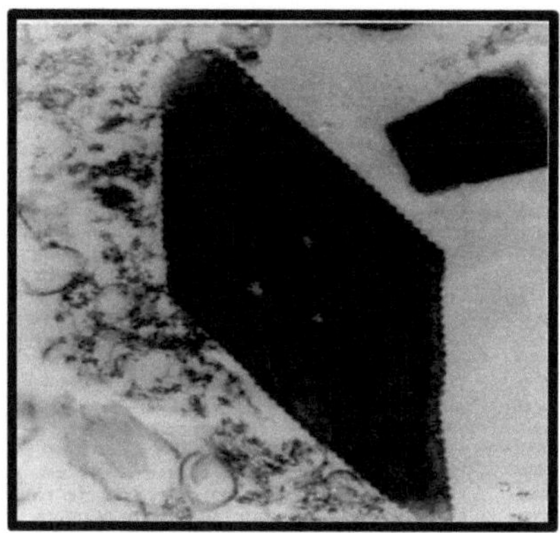

Figure8: cut ultrafine of a Crystal protein Bacillus thuringiensis (JEROME and al , 2002).

4-5-Mode action

According to (JEROME et al , 2002) the entompathogenic activity of this germ is linked to the presence of this parasporal inclusion (crystal) made up of protoxins also called delta –endotoxins. THE crystals have the more often and according to the strains a larvicidal activity on different species of insects belonging to three orders: Pidoptera, Colioptera and Diptera. The crystals synthesized by the bacteria are made up of protoxins, which once ingested by the insect, are digested in an alkaline environment by digestive proteases and transformed into active polypeptide toxins. Delta-endotoxins activated by insect proteases bind to specific receptors located on the cells of the intestinal epithelium, intoxication manifests itself very quickly by significant lesions in the intestine and by paralysis of the digestive tract, leading to an immediate cessation of activity power supply. The death of the insect intervenes in 24 has 48 hours After ingestion crystals and maybe or not accompanied by sepsis, the molecular aspects of the mechanism which lead to the death of insects are not yet clearly defined.

4-6-Advantage And disadvantages of use of the biopesticides Advantage

According to CHARLE And CODERRE (1992), use of the biopesticides offer spectacular short-term benefits:

1. He is of a technology simple And GOOD adapted has the economy global.
2. This is the method Who offers the more of solution real and sustainable mainly due to automatism, the variety of specificity of intrinsic accounting with nature and their ability to evolve with and without direct human intervention.
3. THE percentage of mortality East A criteria of efficiency of this method.
4. A big specificity action and so in principle one lesser hazard For human health and environmental integrity.
5. These are relatively safe biopesticides that are harmless to vertebrates, plants they same, environment And THE pests No targeted.
6. THE insecticides microbial can be easily products in big quantity by traditional fermentation methods.

Disadvantages

According to CHARLE and CODERRE (1992), there specificity of the biopesticides has base of Bacillus thuringiensis gives them economic advantages compared to traditional insecticides, the potential market being smaller for these products:

1. THE spectrum action East narrow For a below- species given bacterial .
2. There toxin East inactivated by there light ultraviolet.
3. He exist a possibility of resistance of the insects has the insecticide microbial.
4. THE toxins insecticides must be ingested For be effective.
5. THE lack of knowledge fundamental on the biology And epidemiology of the micro-organisms concerned.
6. The impossibility of produce economically in mass several kinds potentially useful pathogens.
7. The uncertainty to get of levels students of mortality at the house of THE pests targeted, linked to the impossibility of controlling all the factors which determine the success of the infection, the development of the disease and the decline of the host.

5. Insecticide THE Difluobenzuron : No commercial DIMILIN

5.1. Profile Technical

Specific: lepidopteran caterpillars Double action: larvae and eggs Effectiveness : effective has very weak dose Long-lasting effect

5.2. Identity record

Formulation : suspension concentrated (Sc) Composition: Difluobenzuron

Family : Benzoyl ureas

Ranking : toxicological : No classified Origin: Dutch (CROMPTON)

Presentation: colorless crystals

Solubility : in water : 0.2mg/L has 20 C°.(reference,)

5.3. Formula chemical difluobenzuron

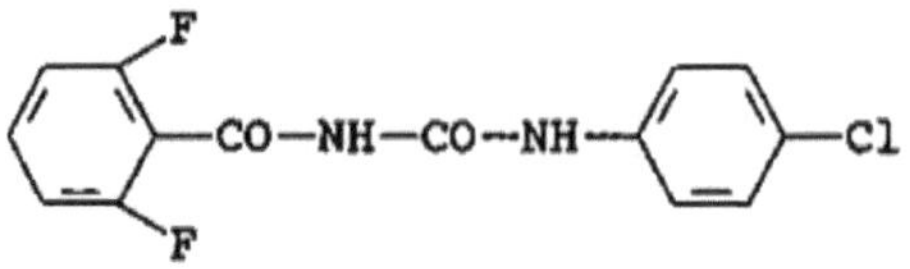

(reference)

5.4. Fashion action

DIMILIN FLO It is A regulator of growth of the insects, present double action

➢ Larvicide : by ingestion on the larvae of the lepidoptera (pyrale of the dates).It blocks the shedding cycle; their activity is manifested during the moulting of the insect. The synthesis of chitin is disrupted, causing serious damage to the endocuticular tissue. .The larvae, which born are not killed Or immediately paralyzed , die at moment of the following moult, the cuticle cannot resist the muscular tension and turgor during moulting.

He n / A not Or little action about insects adults .Dimilin can applied before pupation for get the best results possible, it is recommended to apply it during (1st to 3rd) larval stage, when larval mortality will be higher. If we used on At 4th stadiums larvae, we THE will prevent of se transform in pupae.

➢ Ovicidal of contact : he destroys the eggs filed After THE treatment on THE leaves Or the fruits . He is stable on the foliage, its degradation in the soil varies depending on the organic matter content.

Dimilin has no effect on auxiliary insects which do not consume treated plants, is poorly leachable. Can be applied from spring to autumn. For optimal effectiveness, apply Dimilin in preventive or the appearance of the firsts caterpillars, born must not be applied to the stadiums relatives of marketing because this product can dirty the flowers or foliage. (reference)

5.5. How to use

- Preparation of there porridge :
- GOOD shake Before opening For ensure a total discount in suspension.
- Dimilin FLO is used in spray After dilution In the water.
- Fill THE reservoir has 1/2 L of water.
- Put below hustle And apply THE treatment.

CHAPTER III
MATERIAL AND METHOD

Introduction

The experiment understand two parts GOOD distinct: has know A Breeding of the moth of dates by there presence of dates wormy and the study of the influence of three products on the larval stage (L 3) by ingestion.

1. Material

- For there realization of breeding we has used THE material following:
- Dates wormy in quantity massive.
- Movie of plastic.
- Bottles For recovery THE adults For mating.
- THE tulle in mouslime has mesh fine.
- THE boxes.
- Medium artificial (wheat +date) crushed.
- The water.
- Resistance electric.
- Balance sensitive.

Figure 9 : Material of experimentation

2. THE terms artificial

- Temperature of the order of 28 C°.

- Humidity included between (60- 75%) assured by A water tray .

- Photoperiod from16 hours clear And 8 hours dark.

3. Breeding method

To breed the date moth we have to have a massive amount of dates (varieties Deglet –Nour), these last are in origin of the palm groves of (Sidi Okba, and Tolga). THE dates have summer spread on a movie of plastic In a small bedroom with the artificial conditions to be determined, namely an ambient temperature of around 28°C ensured by an electrical resistance, and the relative humidity of around 70% insured by a water tank permanently within the breeding chamber.

From the appearance THE adults of moth dates we has to proceed to there harvest of the adults garlic moth. The collection of these is carried out in plastic bottles close with fine mesh muslin tulle. After 48 hours adults put in captivity will begin their mating within the bottles. After 48 hours we has proceed has there recovery of the eggs in the boxes plastics of which they were initially put.Once the eggs recover these last were put In of the boxes in tightly closed plastic containing the artificial medium which consists of wheat and crushed dates and put in a temperature has the oven settled has 60°C in order to of destroy all shapes parasitic and mites, the mixture was lightly soaked with water to provide relative humidity favorable for larval development. After three weeks of incubation THE eggs have emerged of the larvae of different stadiums, we This Who We concern We have taken in account THE stadiums L3. THE choice of stadium L3 justify oneself by se sensitivity to treatments.Larvae of stadium L3 was determined based on head capsule size.

4. Technical of treatment

The purpose of this essay consists of determining and comparing,The effectiveness of an entomopathogen only the BT or Bacillus thuriengiensis , associated with Dextrin and a insecticide regulator of growth Diflubenzuron, these last have been used has three doses different vis à vis larvae of the date moth.

4.1. Treatment by Bacillus thuringiensis

This product is used in the form of powder marketed on microlepidoptera of a strain of BT Or Bacillus thuringiensis .THE doses have summer chosen has leave of the approved dose (50g/hl). 1st ^{dose}= 0.25 g. 2nd dose = 0.50 g. 3rd dose = 0.75 g. THE powders of BT have summer weighing by a sensitive balance, each dose has summer mix with water, and according to the product instructions, for our case 0.5 liters of liquid product, we We took distilled water to avoid any disturbances that could influence mortality.

4.2. Treatment by Bacillus thuringiensis with sugar (Destrin)

We used the same technique previously mentioned, as well as the same doses, except in this case we added 5% dextrin sugar. Each dose of BT has been adjusted with 5% of dextrin Then added with some water distilled either a suspension of 0.50 L was considered sufficient for such treatment.

4.3. Treatment by A insecticide regulator of growth THE Diflubenzuron

THE choice of product is justified by the fact that he is already approved on Borer of the dates has a dose of 40 g/hl. The choice of doses. A total of three doses were chosen.

1st ^{dose} = 0.2 g. 2nd dose = 0.4 g. 3rd ^{dose} = 0.6 g

4.4. Preparation larvae destiny to treatment

From mass breeding we have selected the L3 larval stages in sufficient quantity in order to of allow their infestations artificial In THE boxes already having undergoes prior treatment with the products tested for insecticide and BT.

4.5. Mode of treatment

In of the boxes cylindrical in plastic covered of a lid perforated help of a pin to ensure aeration, the food substrate was weighed at a rate of 30 g per box and already sprayed by each dose of each product. In total 05 larvae per box to avoid the phenomenon of cannibalism and at a rate of 03 repetitions per dose).

THE boxes are bets after the treatment In A place obscure.

5. Farms of the data

5.1. Correction of there mortality

THE percentage of mortality observed at the larvae processed is estimated by applying the following formula: Mortality observed =Number of the people dead* 100 / Number total individuals. Mortality observed and then corrected by the ABBOT 1925 formula.

$$MC = \frac{M_2 - M_1}{100 - M_1} * 100$$

M 1 : percentage of there mortality In THE witnesses. M 2 : percentage of mortality in the treated. MC: percentage of corrected mortality.

5.2. Analysis of the variance

According to Dagnelie 1975, analysis of the variance of a series statistical Or of a distribution frequency is the average arithmetic of the square of the gap kind by report has the average. The threshold 5% is adopted in the present study in order to confirm the effectiveness of treatments and doses. For this step we used the STATITCF software.

CHAPTER IV
RESULTS AND DISCUSSIONS

1. Effect of bacillus thuringiensis on stadium larval (L3) of Ectomyelois ceratoniae .

In this part we to study efficiency of a strain bacterial Bacillus thuringiensis screw has screw of the larvae of there moth of the dates In conditions of laboratory, the app of Organic pesticide has given of the mortalities daily Who are instructions In THE painting At Below :

Painting 2: Mortality daily, mortality average And mortality corrected cumulative in percentage of L3 larval stage treated by Bacillus thuringiensis

	DAYS	1	2	3	4	5	6	7
Dose1	T	5	8	9	9	10	11	11
	R1	10	17	18	19	25	25	31
	R2	8	11	18	20	25	27	33
	R3	7	13	17	22	22	26	29
	M	7,6	12,25	15,5	17,5	20,5	22,5	29
	M%	16,66	27,33	35,33	40 ,66	48	52	62
	MC%	12,27	21,01	28,93	34,79	42,22	46,06	57,3
Dose2	T	0	3	7	9	10	12	12
	R1	16	21	25	29	29	30	33
	R2	17	24	23	28	29	31	33
	R3	13	21	24	26	31	31	33
	M	11,5	17,25	19,75	23	24,75	26	27,75
	M%	30,66	44	48	55,33	59,33	61,33	66
	MC%	30,66	42,26	44,08	50,91	54,81	56,05	61,36
	T	2	9	9	10	11	12	13
	R1	20	23	27	30	31	34	35
	R2	15	23	27	30	31	34	36
	R3	15	25	29	30	33	34	36

Dose3	M	13	20	23	25	26,5	28,5	30
	M%	33,33	47,33	55,33	60	63,33	68	71,33
	MC%	31,96	42,12	50,91	55,55	58,79	63,63	67,04

The study of there mortality corrected provoked by there dose 1 of Bacillus thuringiensis watch A rate of low mortality during the first days is increasing At fur and to measure to reach 57.3% on the 7th ^{day}.

Mortality corrected caused by there dose 2 of the entomopathogen tested watch A rate relatively high mortality compared to the previous one, it varied between 30.66 to 44.08% in THE 4th day reach 55.33% and increases gradually 66% In THE 7th day. In indeed the examination of corrected mortality caused by dose 3 of Bacillus thuringiensis shows a relatively high mortality rate compared to doses 1 and 2. It gradually increases to reach 67.4% in 7 th day. THE painting 2 watch Also that he exist A dose effect very remarkable, in effect the mortalities increase according to the doses affected.

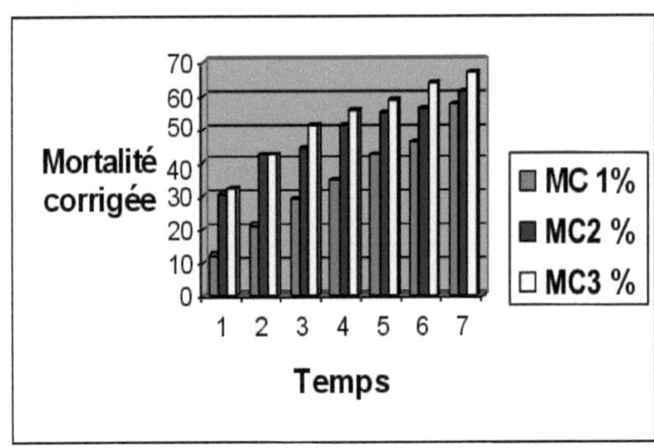

Figure10 : Mortality Daily corrected of stadium L3 of E Tomyelois ceratoniae by Bacillus thuringiensis

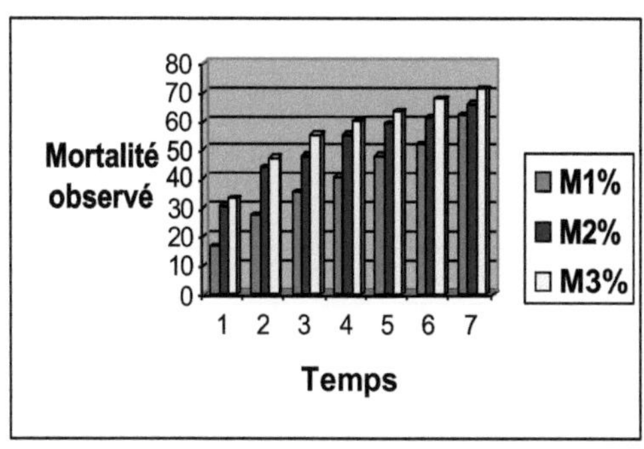

Figure11 : Mortality Daily observed of stadium L3 of ec tomyelois ceratoniae By Bacillus thuringiensis

1.1. Analysis of variance

Table3 : Analysis of variance of the effect of Bacillus thuringiensis on stadium larval L 3 of Ectemyelois ceratoniae

	SCE	DDL	Square AVERAGE	Test F	PROBA	AND	resume	Result
VAR : TOTAL	3366.22	62	54.29					
VAR : F1	771.46	2	385.73	171.13	0.0000			T. H. S
VAR: F2	2448.22	6	408.04	181.03	0.0000			T H.S
WAS : inter F1. F2	51.87	12	4.32	1.92	0.05970			NS
WAS : Residual1	94.67	42	2.25			1.5	6.0%	

Analysis of variance of the effect of Bacillus thuringiensis on larval stage L 3 of Ectomyleois ceratonia highlights the postman dose F1 Who exercises a efficiency very highly significant with probability of 0.0000. However, the analysis of the variance For the postman time East very highly significant Or the probability 0.0000 but for the interaction F1.F2 shows that there is an effect no significant of the order 0.05970.

1.2.Test of NEWMAN And KEULS

Table4 : Ranking of averages For THE factor 1 : dose

F1	LABELS	AVERAGES	GROUPS HOMOGENEOUS
3	D3	2.48	HAS
2	D2	26.05	B
1	D1	20.14	VS

According to THE test of NEWMAN-KEULS, exists three groups homogeneous, THE band HAS represented there dose 3, THE band B represented there dose 2 and The band VS Who represented there dose 1 Or there dose 3 more effective that there dose 2 and 1. This Who permit of THE to classify by degree of efficiency.

Table5 : Ranking averages For THE factor2: time

F2	LABELS	AVERAGES	GROUPS HOMOGENEOUS
7	J7	33.22	HAS
6	J6	30.22	B
5	J5	28.44	VS
4	J4	26.00	D
3	D3	23.11	E
2	J2	19.78	F
1	D1	13.44	G

According to table this After, he exist seven groups homogeneous, THE last day brand there high average mortality is (33.22%) but the first day marks the very low average mortality (13.44%).

2. Effect of Bacillus thuringiensis partner has 5% Dextrin on L3 larval stage.

In this part, three doses of Bacillus thuringiensis associates has 5% Dextrin have been tested on date moth larvae (L3).

Painting 6 : results global of treatment by Bacillus thuringiensis partner has 5% Dextrin.

	DAYS	1	2	3	4	5	6	7
	T	5	8	9	9	10	11	11
	R1	10	19	19	21	23	30	33
	R2	16	16	20	25	27	30	33
	R3	11	21	23	24	27	30	32
	M	8	16	17,75	19,75	21,75	25,25	27,25
Dose1	M%	18	37,33	41,33	46,33	51,33	60	65,33
	MC%	13,68	31,88	35,52	41,38	45,92	55,05	61,04
	T	0	3	7	9	10	12	12
	R1	14	19	24	27	30	34	37
	R2	17	21	24	26	28	32	34
	R3	19	20	24	26	28	32	34
	M	12,5	15,75	19,75	22	24	27,5	29,25
Dose2	M%	33,33	40	48	52,66	57,33	65,33	70
	MC%	33,33	38,14	44,08	47,97	52,58	60,6	65,9
	T	2	9	9	10	11	12	13
	R1	16	22	25	27	30	35	40
	R2	19	22	25	27	30	36	38
	R3	19	21	25	26	31	35	40
	M	14	18,5	21	22,5	25,5	29,5	32,75
Dose3	M%	36	43,33	50	53,33	60,66	70,66	78,66
	MC%	34,69	37,72	45,05	48,14	55,79	66,65	75,47

The exam of the mortalities corrected caused by there dose 1 of Bacillus thuringiensis partner has 5% Dextrin shows a mortality of the order of 13.33% in the first day and increases gradually to reach 61.04% in the 7 th day. The exam of the results of mortality corrected cause by there dose 2 watch that A mortality

33.33% In THE first day And increase At fur has measure For reach 65.9% in the 7th [day]. THE rate of corrected mortality checked in For there dose 2 is more important by relation to there dose 1. Thus the corrected mortality rate in dose 3 remains the highest with a mortality rate of around 75.47%, this rate remains higher compared to doses 1 and 2

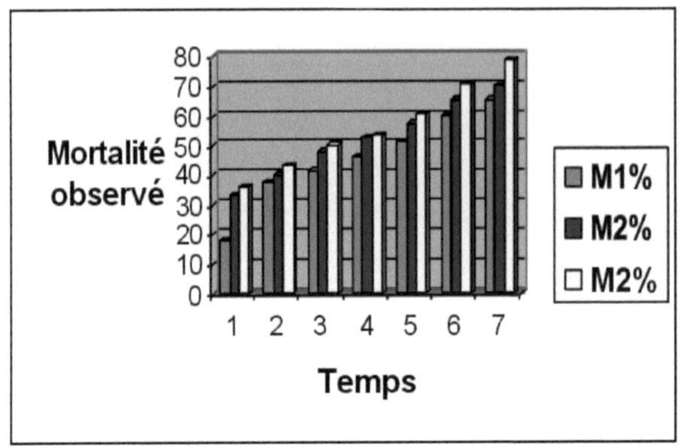

Figure12 : Mortality Daily observed of stadium L3 of Ectomyelois ceratoniae by Bacillus thuringiensis partner has 5% Dextrin

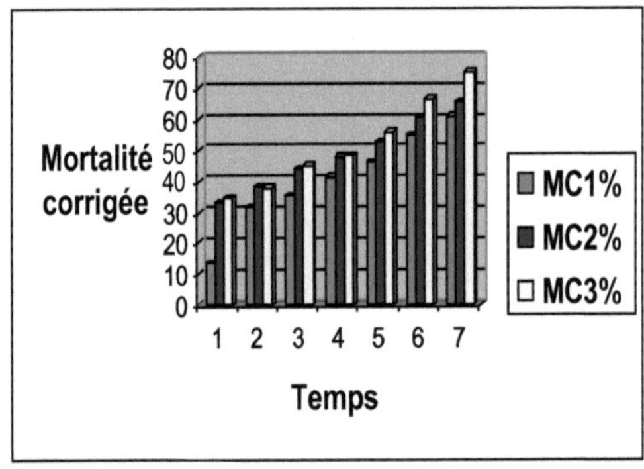

Figure 13: daily mortality corrected of stadium L3 of Ectomyelois ceratoniae by Bacillus thuringiensis associated with 5% Dextrin

2.1. Analysis of variance

Table7 : Analysis of the variance of the effect of Bacillus thuringiensis partner has 5% Dextrin on L3 larval stage.

	SCE	DDL	Square AVERAGE	Test F	PROBA	AND	resume	Result
VAR : TOTAL	2991.71	62	48.25					
WHERE : F1	236.86	2	118.43	49.74	0.0000			THS
WHERE: F2	2629.49	6	438.25	184.06	0.0000			THS
WHERE : inter F1. F2	25.36	12	2.11	0.89	0.5661			NS
VAR : Residual1	100.00	42	2.38			1.54	6. 0%	

The analysis of variance shows that the dose and time factors are very highly significant. With a probability of 0.0000 but the interaction F1 .F2 shows there has a non-significant effect i.e. a probability of order 0.5661.

2.2. Test of NEWMAN And KEULS

Painting 8: Ranking averages For THE postman 1 : dose

F1	LABELS	AVERAGES	GROUPS HOMOGENEOUS
3	D3	28.05	HAS
2	D2	26.19	B
1	D1	23.33	VS

This test allows you to constitute three groups homogeneous. The efficient East very net For THE band To which represents dose 3.

Table9 : Ranking of averages For THE postman 2 : time

F2	LABELS	AVERAGES	GROUPS HOMOGENEOUS
7	J7	35.67	HAS
6	J6	32. 67	B
5	J5	28.22	C
4	J4	25.44	D
3	J3	23.22	E
2	J2	20.11	F
1	J1	15.67	G

The exam of painting above shows the existence of seven groups homogeneous, the toxicity is revealed by the group HAS Who represents the last day when we recorded a mortality rate of 35.67% on the other hand the first day was represented by the last group G where the recorded mortality rate is the lowest, i.e. 15.67%.

3. Effect of regulator of growth Diflubenzeron.

In this part ,we study efficiency of three doses of a Diflubenzeron regulator with respect to the larvae (L3) of Ectomyelois ceratoniae Zeller. After treatments, the results are recorded in the following table.

Table 10: Results global of treatment by Diflubenzeron.

	DAYS	1	2	3	4	5	6	7
Dose 1	T	5	8	9	9	10	11	11
	R1	4	5	8	11	14	18	20
	R2	2	5	8	12	14	18	20
	R3	1	6	8	9	12	18	20
	M	3	6	8,25	10,25	12,5	16,25	17,75
	M%	4,66	10,66	16	21,33	26,66	36	40
	MC%	1,02	2,89	7,69	13,54	18,51	28,08	32,58
Dose 2	T	0	3	7	9	10	12	12
	R1	6	8	12	16	17	23	25
	R2	2	6	12	16	17	23	25
	R3	2	6	10	16	16	20	22
	M	2,5	5,75	10,25	14,25	15	12,5	21
	M%	6,66	13,33	22,66	32	33,33	44	48
	MC%	6,66	10,64	16,83	25,27	25,59	36,36	40,9
Dose 3	T	2	9	9	10	11	12	13
	R1	7	8	14	19	20	24	26
	R2	4	8	14	15	18	24	26
	R3	2	7	13	14	16	24	26
	M	3,75	8	12,5	14,5	16,25	21	22,75
	M%	8,66	15,33	27,33	32	36	48	52
	MC%	6,79	6,95	20,14	24,44	28,08	40,9	44,82

Examination of the table below shows that the corrected mortality caused by dose 1 of Diflubenzeron is very weak in the 3 first days (1.02%, 2.89%, 7.69%), it ends gradually increasing to reach 32.58% For the 7th day . Corrected mortality caused by dose 2 in the first is of the order of 6.66%, the latter increases slightly from 2nd day to reach 10.64%, from the 7th day mortalities reported reached rate of the order by 40.9%, he is at point out that only the dose 3 remains more effective compared to doses 1 and 2 her rate reached 44.82%.

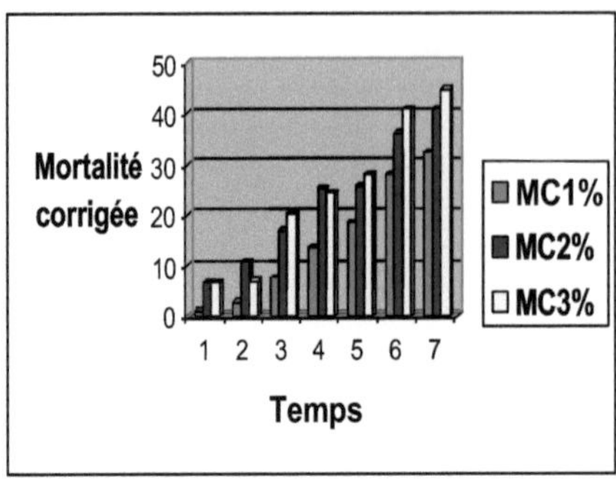

Figure 14: Mortality Daily corrected of stadium L3 of Ectomyleois ceratoniae provoked by THE regulator of growth THE Diflubenzeron

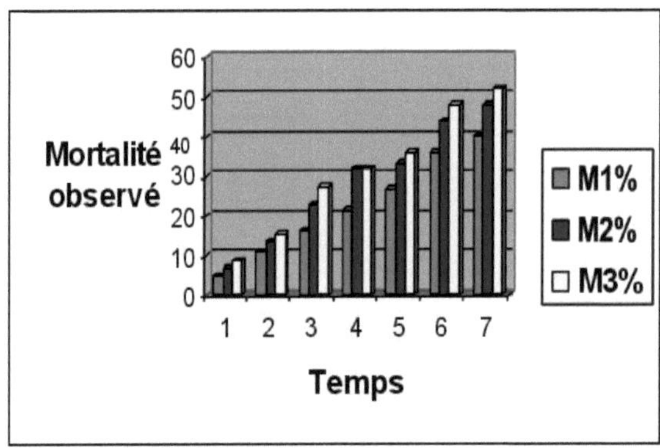

Figure 15: Mortality Daily observed of stadium L3 of Ectomyleois ceratonia by regulator of growth Diflubenzeron.

3.1.Analysis of variance

Painting 11: Analysis of there variance of efficiency of Diflubenzeron on THE on stadium larval L3.

	SCE	DDL	Square AVERAGE	Test F	PROBA	E, T	Resume	Result
WHERE : TOTAL	3245.65	62	52.3 5					
WHERE : F1	230.89	2	115.44	62.7	0.0000			THS
WHERE: F2	2901.87	6	483.65	262.67	0.0000			THS
WHERE : inter F1.F2	35.56	12	2.96	1.61	0.1257			
VAR : Residual1	77.33	42	1.84			1.36	9.9%	

THE painting 11 show that THE factors dose and time are very highly significant With a probability of the order of 0.0000 but the interaction F1 .F2 shows that he y has a non-significant effect and remains relatively greater than 0.05.

3.2. Test of NEWMAN And KEULS

Table 12 : Ranking of the averages For THE postman 1 : dose

F1	LABELS	AVERAGES	GROUPS HOMOGENEOUS
3	D3	15.67	HAS
2	D2	14.29	B
1	D1	11.1	VS

Three homogeneous groups HAS, B, vs. The exam of Test Newman does came out there dose 3 represented by the band HAS is here more effective that the dose 2 and finally there dose 1 remains the least effective

Painting 13 : Ranking of averages For the factor2 : time

F2	LABELS	AVERAGES	GROUPS HOMOGENEOUS
7	J7	23.33	HAS
6	J6	21.33	B
5	J5	16	VS
4	J4	14.22	D
3	J3	11	E
2	J2	6.56	F
1	J1	3 .3 3	G

The averages reveal seven homogeneous groups, the first group A represents the last day effective that other groups Or he represented 23.33% mortality.

Discussion

During this work essentially devoted to the study and evaluation of the toxicity of a strain commercialized of a entomopathogenic Bacillus thuriengiensis alone, Then added to 5% Dextrin and a growth regulator Diflubenzuron on stage L3 of a date pest Ectomyelois ceratoniae.

Bacillus thurigiensis bacterial solution treatment trial in three different doses noted previously the mortalities corrected more important in the corresponding dose 3 (67.04%) with the 7th day. We notes that mortalities corresponding at the dose: D3 are superiors has those of D1 and D2. Our observations daily We have permit of to remark that firsts mortalities appear from the first day after treatment with the bacteria. However, the bacterial strain Bactospeine, commercial product based on Bacillus thuriengiensis , its shown enough effective catchy a mortality larval going up to 68%, these are in agreement with the results obtained by (JARARAYA 2003). DHOUIBI (1992) notes that the real impact of this product on dates remains relatively low, and note that THE rate of fruits crooked only decreasing of 30% approximately by report At witness, and explains for his part that the relative effectiveness is due both to the method of application (aerial treatment) and to the feeding behavior of the caterpillar, taking into account that it leads a mainly endophytic life, is protected from any contact with the toxin. For its part (ABDEL-RAZEK, 1998) confirms that formulations based on Bacillus thuriengiensis var Indiana And var morrisoni are shown very effective with regard to of a stored foodstuff microlepidopetra Cadra cautella. Indeed, although the mortality rate recorded by this bacteria seems to us to be of little importance compared to what we expectations. Account of the mode of application which was the spraying of the food product, we believe that only the soaked parts retained the spores entompathogens, as well the larvae ingested a part of substrate eating Who did not contain spores, these is comparable with the work of (SCALO et al, 1997) which showed that the mortality rate is related to the mode of application , Thus they observe that on clusters pulverized of the rate mortality less important by relation to of the clusters completely soaked, and these for the leafrollers of the cluster, a lepidopteran close to our insect. As for the test of toxicity of this entomopathogenic partner has 5% of Dextrin THE results have watch This Who follows. Mortality In there dose 1 match (61.04%), there dose 2 (65.9%) but there dose 3 more important with mortality 75.74%. The exam of the results mentioned this -After reveal that the addition of Dextrin effectively influences the mortality rate of larvae, it turns out that the moth has the power of ingestion a lot more important in assimilating THE Dextrin sugar This Who induces a severe toxicity, our results are similar to those of (SCALO et al, 1997). The latter showed that the addition of 1% of Cyclodextrin type sugar to the treatment mixture clearly increased the effectiveness of certain strains of Bacillus , particularly on a Cochylis leafroller, it should be noted that certain strains tested by these authors mortality reached 100 %. The toxicity of a growth regulator Diflubenzuron shows relatively low mortality rates in three doses different ; D1, D2 and D3, on THE larvae of the moth (L3). compared to those of the entomopathogen tested. Our observations showed that there toxicity of Diflubenzuron stay significantly lower to those Bacillus s tested alone and in addition with Dextrin. Knowing that it does not have a shock effect (ALLACHE, com, Pers), certainly with the mode of action of this growth regulator, the latter blocks moulting from the larval stage to the nymph. Account tenuous of non availability of the works of This regulator of growth against the date moth, we focused on comparing our data with that of the literature on other pests. THE fashion action of these regulators alter considerably in reducing THE content in lipids, proteins And in carbohydrates has leave of 4th day ingestion, (BOUDEBOUS AND DJOUMEH, 1995) demonstrated this effect on the females adults of Tenebrio molitor treated by ingestion of a dose of 10mg/g, of her side (SOLTANI-MAZOUNI, 1994) observed the same alterations on pupae of this same beetle. Other studies conducted on lepidoptera Lymantria dispar treated by ingestion at stage L3 at a rate of 500 mg/l (OUAKID, 1991), similar observations were observed in Cydia pomonella by application of a dose of 0.5 mg of Diflubenzuron to young pupae By elsewhere, he seems that these compounds affect the metabolisms ovaries, they are analogues of juvenile hormone (SOLTANI, 1998).

CONCLUSION GENERAL

In THE frame of This work we adopted has the study efficiency biopesticide has base of Bacillus thuringiensis alone, Bacillus thuringiensis partner with 5% of sugar Dextrin and the insecticide Difluobenzeron on larval stage of date moths Ectomyelois ceratoniae .THE results Who We have obtained as following : Efficiency of Bacillus thuringiensis on the stadium larval L3 watch a mortality highly significant regarding the three doses but dose 3 more effective than the other doses.THE rate of mortality In there dose 3 (67.04%) but at the dose 2 (61.63%) and at the dose 1 (57.3%). Efficiency of Bacillus thuringiensis partner with sugar Dextrin 5% East more effective that Bacillus thuringiensis alone. Mortality in the dose 3 (75.47%) and at the dose 2 (65.9%) and the dose 1 (61.04%). Difluobenzeron possesses a efficiency more lesser that Bacillus thuringiensis alone and Bacillus thuringiensis associated with 5% Dextrin. THE rate of mortality In there dose 3 (44.82%) but at the dose 2 (40.9%) and at the dose 1 (32.58%). He We appeared in OUR work that the fight biological by Bacillus thuringiensis possesses greater effectiveness than chemical control in eliminating the harmfulness of date moth larvae Ectomyelois ceratoniae.

SUMMARY

Printed by Books on Demand GmbH, Norderstedt / Germany